AF270584

OPOSSUMS

by Elizabeth Andrews

Cody Koala

An Imprint of Pop!
popbooksonline.com

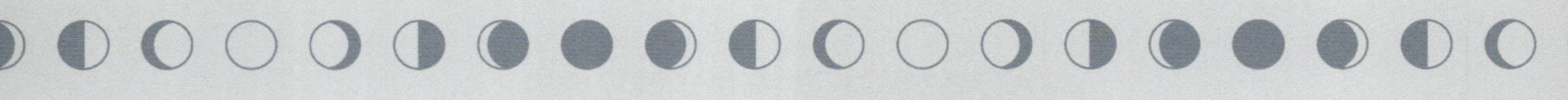

abdobooks.com

Published by Pop!, a division of ABDO, PO Box 398166, Minneapolis, Minnesota 55439. Copyright ©2023 by Abdo Consulting Group, Inc. International copyrights reserved in all countries. No part of this book may be reproduced in any form without written permission from the publisher. Cody Koala™ is a trademark and logo of Pop!.

Printed in the United States of America, North Mankato, Minnesota

052022
092022

THIS BOOK CONTAINS RECYCLED MATERIALS

Cover Photo: Shutterstock Images
Interior Photos: Shutterstock Images, Rolf Nussbaumer Photography/Alamy Stock Photo

Editor: Grace Hansen
Series Designer: Laura Graphenteen

Library of Congress Control Number: 2021951852
Publisher's Cataloging-in-Publication Data
Names: Andrews, Elizabeth, author.
Title: Opossums / by Elizabeth Andrews
Description: Minneapolis, Minnesota : Pop, 2023 | Series: Twilight Animals | Includes online resources and index
Identifiers: ISBN 9781098242091 (lib. bdg.) | ISBN 9781098242794 (ebook)
Subjects: LCSH: Opossums--Juvenile literature. | Marsupials--Behavior--Juvenile literature. | Twilight--Juvenile literature. | Nocturnal animals--Juvenile literature. | Nocturnal animals--Behavior--Juvenile literature. | Marsupials--Juvenile literature.
Classification: DDC 591.518--dc23

Hello! My name is

Cody Koala

Pop open this book and you'll find QR codes like this one, loaded with information, so you can learn even more!

Scan this code* and others like it while you read, or visit the website below to make this book pop.

popbooksonline.com/opossums

*Scanning QR codes requires a web-enabled smart device with a QR code reader app and a camera.

Table of Contents

Chapter 1

What's on the Menu?

Opossums are hungry little **omnivores**. They eat almost anything, like insects, garden vegetables, trash, and roadkill. Scientists sometimes call opossums "nature's clean-up crew."

5

Snake bites do not hurt
opossums even if they
are poisonous.

Opossums are **hardy** creatures who start moving at twilight. They eat things that most animals try to avoid, like poisonous snakes. They eat so many ticks that it helps protect other animals from sicknesses the bugs carry.

Awesome Opossums

Opossums are about the size of a house cat. They have **coarse** fur that ranges from white to black. The paws, ears, and tail are naked. Their heads are long and pointed with lots of teeth.

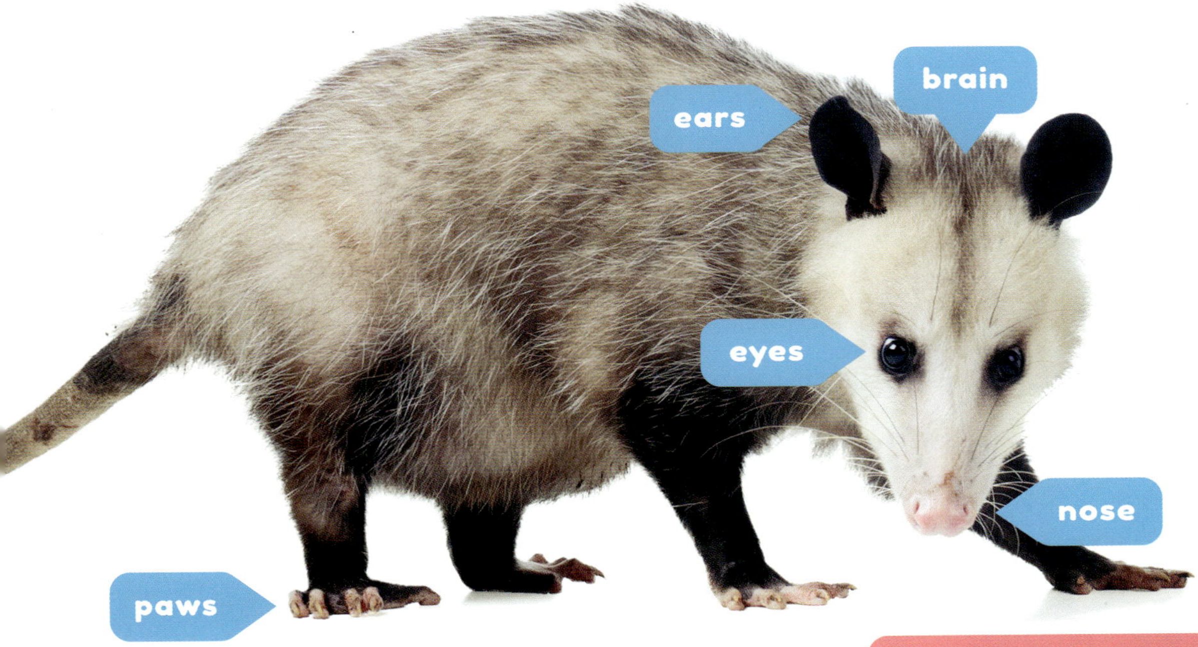

Paws: each has five toes to help with climbing

Brain: great memory to remember where food is

Ears: good hearing to stay alert for **predators**

Eyes: bad eyesight makes them rely on other senses

Nose: strong sense of smell for hunting food

Baby opossums
can hang from tree
branches by their tails.

Opossums have a tail that acts like another **limb**. Opossums use it to hold small objects. Their sharp claws make them expert tree climbers.

Opossums do not make their own homes. They choose different places to sleep day-to-day. An empty squirrel nest, a cozy space in a wood pile, or an old shed all work just fine.

Chapter 3

Playing Dead

Opossums usually try to run from their **predators**, but they are not very fast. When they can't get away, opossums get so scared they collapse in **shock**. People say it looks like they're playing dead.

Explore links here!

Their heart rate slows.
They foam at the mouth and
their tongue flops out. An
opossum will also **secrete**
bad smells. It doesn't taste
good to a predator.

Chapter 4

Marsupial Mamma

Opossums are marsupials. Babies are born before they are fully **developed**. They are carried in a pouch on their mother's belly where they can grow more. Babies drink their mother's milk.

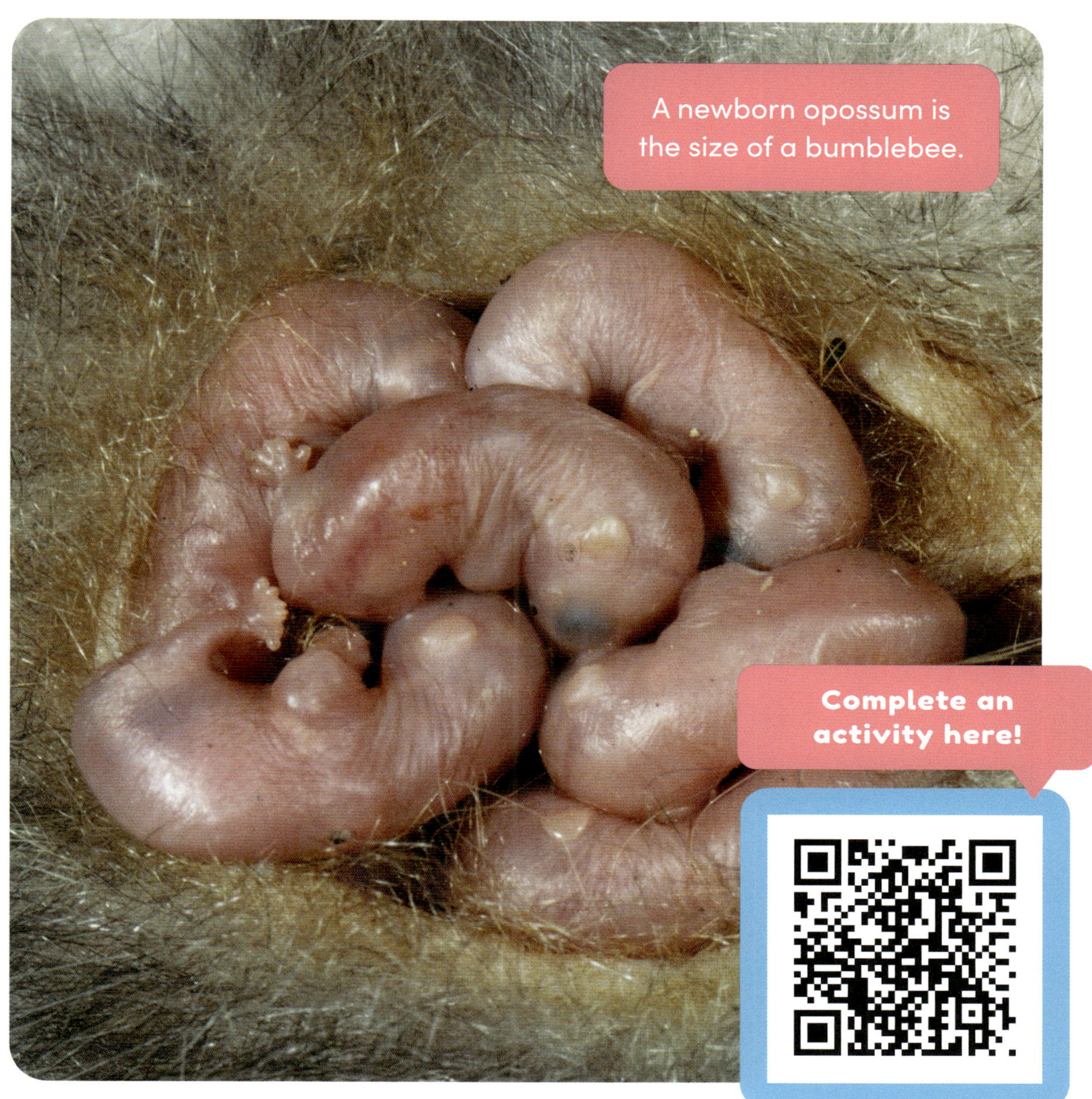
A newborn opossum is the size of a bumblebee.
Complete an activity here!

Female opossums can give birth to up to 20 babies at once. Usually only eight survive to live outside of the

pouch. Sometimes babies
ride on their mother's back
when they get too big for
the pouch.

Making Connections

Text-to-Self

Did you know anything about opossums before reading this book? What new things did you learn?

Text-to-Text

Have you read any other books about marsupials? If so, how were those animals similar to and different from the opossum?

Text-to-World

Do you think opossums are important animals to have in neighborhoods? Explain your answer.

Glossary

coarse – rough texture.

developed – having a brain and body healthy enough to survive outside of the womb.

hardy – tough and brave.

limb – a part of the body that can move and bend.

omnivore – an animal that eats plants and other animals.

predator – an animal that hunts other animals for food.

secrete – to release fluid or other substances out of the body.

shock – a dangerous weakening of the body.

Index

Online Resources

popbooksonline.com

Thanks for reading this Cody Koala book!

Scan this code* and others like it in this book, or visit the website below to make this book pop!

popbooksonline.com/opossums

*Scanning QR codes requires a web-enabled smart device with a QR code reader app and a camera.